THE MYSTERIES OF PLUTO

An Alternative Approach to the Solar System's Most Unique World

by

James Stuart

authorHOUSE®

AuthorHouse™
1663 Liberty Drive, Suite 200
Bloomington, IN 47403
www.authorhouse.com
Phone: 1-800-839-8640

First published by AuthorHouse 10/30/2007

ISBN: 978-1-4343-1622-6 (e)
ISBN: 978-1-4343-1621-9 (sc)

Library of Congress Control Number: 2007905921

Printed in the United States of America
Bloomington, Indiana

This book is printed on acid-free paper.

It was still only about thirty years ago that our knowledge of the Pluto/Charon system had advanced far enough for Carl Sagan to describe the two worlds as being like "Ariel orbiting Triton." As the Hubble Space Telescope, occultation of stars in the foreground, and periodic eclipses of the two worlds by each other advance our knowledge, Dr. Sagan's description is still remarkably prescient and accurate. Even so, to the average person, the comparison of the Pluto/Charon system to a moon of Uranus orbiting a moon of Neptune seems oddly remote.

This need not be the case, however. As of this writing, we know far more about Pluto and its moon than most people can possibly imagine. We know, for example, that Pluto's color is primarily red, and that it has two distinctive polar caps. We know that it has significant winds blowing across its surface, and we know that its surface color varies with latitude. Starting in the south, there is a white polar cap, followed by a band of reddish material in the mid-southern latitudes, followed by a white band near the equator, followed by a reddish band at mid-northern latitudes, and topped by a white northern polar cap. [1] This alternating of five red and white bands, if it were not for other features, would be more reminiscent of the American flag when viewed head-on than the moon in the miniature image of forty-five years ago.

As startling as this imagery is, however, Pluto is a far stranger world than its pictorial description. In fact, Pluto is so bizarre that it is as if nature has played a little joke on us by taking the strangest attribute of almost every other planet in the solar system and combining those attributes in one world. Consider for a moment the following:

Pluto

Like Mercury	is extremely tiny
Like Venus	rotates retrograde
Like Earth	is considered a double-planet system
Like Mars	is a red planet
Like Jupiter	has at least one giant red spot, maybe two

Like Jupiter's moon Europa	may have a significant sub-ice ocean
Like Saturn's moon Enceladus	may be conducive to life
Like Uranus	has an axial tilt of more than ninety degrees
Like Neptune	may have very-high-velocity surface winds
And Like Neptune's moon Triton	may have Nitrogen geysers below −400 deg.

Given this amount of strangeness, it is easy to see why scientists were eager to launch the New Horizons mission. However, even Pluto's many unique features are not the whole story. There are three huge questions about Pluto that it is hoped the New Horizons probe will be able to answer. Is Pluto a planet? Is life possible on Pluto? Can Pluto tell us anything unique about the origins of our own solar system? These are the three questions that we will examine and offer opinions on in the next sections.

Is Pluto a Planet ?

T he current debate over the planetary status of Pluto and other small objects clearly demonstrate the need for a logical framework of small-body classification. Certainly, the demotion of Pluto-sized bodies seems to be in vogue. However, scientists cannot base planetary status on what is currently in style. On the other hand, recent public opinion polls on the demotion of Pluto seem to indicate that the overwhelming majority of respondents believe that science got it wrong in this case. Just as surely as scientists cannot decide status based on the current styles, they cannot decide status based upon public opinion either. Somewhere between the middle of these two extremes, there must lie a compromise solution satisfactory to everyone. Such a solution would have to have the merit of being scientifically objective, however. The purpose of this paper is to put forward just such a solution, and our methodology in this case is to classify nonstellar objects by creating a planetary Hertsprung-Russell diagram.*[1]

A Hertsprung-Russell diagram of stellar objects, it will be remembered, is graphed as mass/luminosity on the y-axis versus temperature/stellar class on the x-axis. Additional complications arise for the x-axis in particular when we remember that the highest-temperature objects are situated in the lower left-hand corner of the graph rather than the lower right-hand corner. In addition, we must remember that the universal basis of the stellar H-R diagram is set on the mass and temperature of our Sun both being equal to one unit.

---

[1] It is believed that this methodology is unique to the author.

In deriving a planetary Hertsprung-Russell diagram, we simply need to find proxies for planetary temperature and planetary mass, have the hottest planetary temperatures chart out to the lower left-hand corner of the diagram, and set Earth mass and temperature proxies to 1. The planetary equivalent of solar temperature we have chosen is the distance from the sun in astronomical units. This works well, since it fulfills the additional requirement that the hottest objects (those closest to their respective stars in this case) fall in the lower left-hand corner of the diagram. Here, it seems that the best scale for this distance measure is based on Log(10) in AU. This seems to be natural, as the object that orbits closest to its own sun that we currently know of circles a red dwarf binary at a radius of only 2 million miles or so, or about 10^{-2} AU, while the objects that orbit farthest from their own suns are naturally cold, brown dwarf-like objects in each star's Oort cloud revolving around their respective suns, sometimes at distances of up to 10^5 AU. With this definition in hand, we now turn our attention to the planetary proxy for solar mass, and for that we choose planetary diameter. Here it is easy to see that this must be a proxy for planetary mass, as we need only multiply by pi/6 and some ratio of planetary density to Earth density in order to come back to the planet's actual mass. The scale we chose here is based on Log(2) of Earth's diameter. The reason for this becomes apparent when we realize that the I.A.U. minimum diameter for roundness and consideration as a planet is 500 miles, or about 1/16 the diameter of the Earth, while the largest objects which can be planetary bodies without actually becoming stars are large brown dwarfs whose absolute upper size limit is around 1/3 of a solar diameter, or about 2^5 Earth diameters. It seems as if these definitions successfully fill all the proxy requirements for creating a planetary H-R diagram, so we will now test our methodology by plugging some numbers and spinning up our first prototype.

Planetary Hertsprung Russell

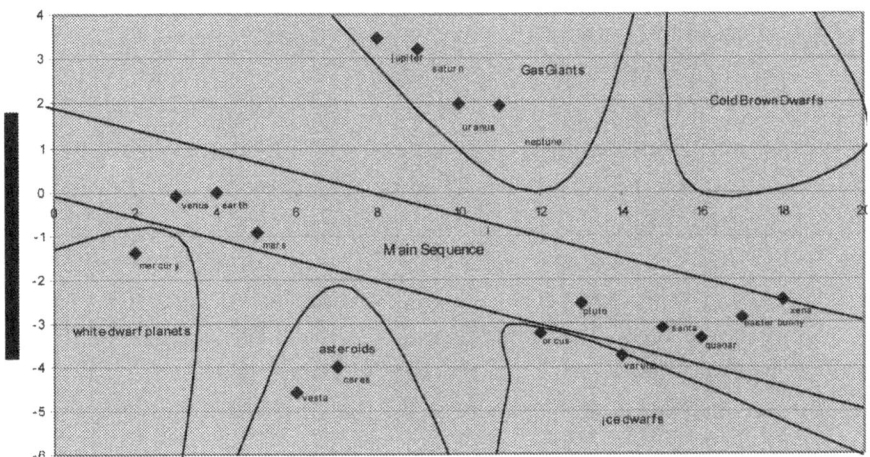

Distance A.U. Log(10) (Proxy for Temperature)

The above diagram clarifies the situation immensely and leads to some interesting conclusions and speculations. First, brown dwarfs that orbit very close to their stars fall in the far upper left-hand corner of our diagram, and they must be the planetary equivalent of class B stellar objects, as they fulfill the same role on the planetary H-R diagram as blue supergiants do on the stellar H-R diagram.

Another interesting feature is that Jupiter-size planets that orbit their stars at distances between 10^{-2} and 10^{-1} AU seem to be analogous to blue giants on the stellar H-R diagram. Since blue giants on the stellar H-R diagram represent a very small percentage of total stellar population, if our analogy holds, then we are forced to the interesting conclusion that Jupiter-size objects relatively close to stars are rather rare as far as planetary objects are concerned. This seems to be saying that the ubiquitousness of these objects in our current extrasolar object surveys seems to be a bias of our measurement abilities rather than a true phenomenon. Another conclusion is that this boosts the probability of extraterrestrial intelligence over current estimates.

Recently, a Neptune-size object seventeen Earth diameters in size was discovered circling a red dwarf star at a distance of about 3 million miles. An interesting test of the planetary H-R diagram is what it tells

3

us about objects such as these. We can check this in a second if we locate 'our' Neptune on the planetary H-R diagram and move it left until it reaches the 10^{-2} AU distance scale. When we do that, we see that Neptune-size objects orbiting close to their suns seem to fall just at the far upper left-hand corner of what appears to be a main sequence on the planetary H-R diagram. The conclusion here seems to be that these objects are probably much more common than we realize.

Continuing on in this fashion, we can see that Mercury-like planets fall in the lower left-hand corner of our diagram, and so, being very hot and very small, fill the same roll on the planetary H-R as white dwarfs do on the stellar H-R diagram. Here again, another interesting analogy takes shape. White dwarf stars tend to be at the end of the evolutionary sequence of objects that were once much larger. These objects usually are red giants that have blown off their outer gas shells during the planetary nebula stage. If Mercury's evolution behaved similarly, then we would conclude that Mercury was once a much larger object and is probably the molten core of an object that had its outer gas shell blown away by the solar wind; Mercury was almost certainly a Jupiter-size gas giant if this is the case; however, Just like their stellar counterparts, Mercury-like planets do not appear to be main sequence.

An interesting feature is the placement of asteroids like Ceres and Vesta on our diagram. They fall in the middle bottom of the planetary H-R diagram. Since no analogous class of stellar objects fills this region on the stellar H-R diagram, the conclusion here seems to be that asteroids fail *all* definitions of planetary status. Indeed, a new definition may now be proposed, and that is, if an object in question falls in the middle bottom of the planetary H-R diagram, it is to be considered an asteroid as opposed to a planet.

Moving on to the lower right-hand corner of our diagram, we encounter objects like Varuna, Orcus, Quaoar, and the Kuiper Belt Objects nicknamed Santa (found near Christmas) and Easter Bunny. These objects appear to fill the same role on the planetary H-R diagram as red dwarfs do on the stellar H-R diagram. Several interesting conclusions can be drawn from this. First, despite their small size, ice dwarf planets may be the equivalent of class M stars on the Stellar H-R diagram, and class M objects are still main sequence, even though some class M objects are so far removed to the lower right-hand corner of the stellar H-R diagram that their classification is tentative at best.

This tends to suggest that these objects are, in a sense, planets, although their classification can be legitimately questioned. The biggest reason for questioning a KBO's classification has to do with loosely constrained estimates of size. If we look at our diagram above, for instance, Varuna and Orcus definitely fall off of the main sequence and therefore are ice dwarfs, even when using their median value diameters. Quaoar, Santa, and Easter Bunny, on the other hand, appear to be on the main sequence and thus are planets, but all three of these objects have a median diameter of about 1,500 km+–300 km, [2] leaving a huge swing in interpretation of the placement of these objects on the diagram, and when more refined estimates of size come in, they may very well fall off of the main sequence and into ice dwarf status. A second point for these objects is that, if seen as planets, just as class M stars are the overwhelming numerical majority of all stars, so too must ice dwarfs be the overwhelming majority of all planetary objects. If our analogy holds, we may even be able to put a number to this majority. Just as red dwarfs constitute 85% [3]of all known stars, ice dwarfs may constitute 85% or more of all known planetary bodies. Another interesting circumstance is that it is thought that about a quarter of all red dwarfs are binaries, meaning that at least 22% of all known stars are red dwarf binary systems. If this analogy carries over, we can expect ¼ of all Kuiper belt ice dwarfs to have moons. The numerical implications of these ideas are staggering. We could expect that there are *at least* sixty to seventy more, and possibly *many* more ice dwarfs in the outer solar system at least the size of Pluto yet to be discovered, all of which could be considered planets in their own right, and furthermore, at least fifteen to twenty of these objects, if not many more, will have moons.

This brings us to the case of Pluto itself, and to a lesser extent, that of Xena/Eris. At 39 AU and a diameter of about 2^{-3} that of Earth, Pluto definitely falls on what would be considered a main sequence on the planetary H-R diagram, as does Xena/Eris, along with Earth, Mars, and Venus. [4] Pluto's position, in fact, is right on the transition point from main sequence to ice dwarf status, and this is undoubtedly emblematic of the confusion surrounding Pluto's true classification. Interestingly enough, Pluto's existence right on this transition point leads to an interesting hypothesis on how it came to be in its current orbit. Since Pluto looks so much like ice dwarf and red dwarf planets like Varuna and Orcus, it is probably reasonable to assume that it was once one of these objects itself, Which means that Pluto did not start out as a moon lost by Neptune, but as an object that was injected into the inner solar system out of the class M objects. The only thing that could do this is something massive, and that suggests that there may exist a massive brown dwarf in either the far Kuiper Belt or a nearby Oort cloud which has yet to be detected. However, the analogous position on a stellar H-R diagram would be near class K, and since class K is definitely main sequence on the stellar H-R diagram, there can be little doubt that Pluto classifies as a true planet in our schematic, especially since its diameter is accurately known. Bolstering this decision is the weakness in the recently accepted definition for planethood stating that a body must dominate its nearby surroundings in order to be considered a planet. The question that arises is, exactly what is it that is meant by *nearby surroundings*? If *nearby surroundings* is defined to be of a class size the size of our solar system, then Jupiter would be the only planet in our solar system by this definition. If, on the other hand, a planet's local neighborhood was to be defined as, say, the inner solar system, then neither Earth nor Venus, being almost the same size, could be said to dominate the terrestrial planets, and so neither would classify as a planet, which is absurd. This can be drawn out to the ultimate extreme by saying that Jupiter is not that much bigger than Saturn and that Saturn is not that much bigger than Uranus or Neptune and that therefore none of the gas giants clearly dominate their neighborhoods either, leaving us with the ultimate absurdity of a solar system which, in a way, has no planets at all by this definition. In addition, Pluto has three moons, none of which have lost their status as moons yet, as far as is known. Which leaves the question, how can a true moon orbit a non-planet? It is our opinion that the concept of the planetary H-R diagram calls for the reinstatement of Pluto as the ninth planet. In addition, Xena/Eris would probably meet the definition of the tenth planet, as its diameter is accurately known and it falls on the main sequence also.

This leaves the gas giants of our solar system yet to be discussed. Their position on the planetary H-R diagram is analogous to the transitional phase between the main sequence and red giants on the stellar H-R diagram. An interesting implication of this classification is that gas giants might have wound up as gas giants because at some point very early on in their evolution they exhausted their original fuel supply, which was relatively short-lived compared to the uranium reactors in the cores of terrestrial planets, and started burning a secondary fuel. The effect of this fuel switch made the cores of the gas giants relatively enriched in the heavier fuel, and thus denser. This was compensated for by the outer accretion disk expanding and cooling, thus creating a very large planet through the interplay between temperature dynamics and the energy needed to bind specific gases rather than metallic elements.

Finally, brown dwarfs would, of course, be the equivalent of red supergiants on our planetary H-R diagram. In fact, the planetary H-R diagram can actually be tacked on to the stellar H-R diagram, as the brown dwarfs of the upper right-hand corner of the planetary H-R diagram must eventually become the red dwarfs of the lower right corner of the stellar H-R diagram as they become more massive. The implication of this is that both planetary and stellar evolution can be unified in one basic process, while another implication is that there may be an undiscovered or undescribed process of planetary evolution, much as their exists a process of stellar evolution. Of course, this has yet to be proven, but it would be fascinating if true.

To sum up, the concept of a planetary H-R classification scheme leads to the following conclusions:

OBJECTS	PROPOSED STATUS
Ceres, Vega	Not planets
Ice Dwarfs	Problematical; probable planets, but questionable
Pluto	Definite planet; needs reinstatement
Xena/Eris	Almost certainly planet ten
Solar System	New planet count would be ten
Small diameter, small AU	Asteroids, middle bottom of H-R classification.

Finally, we come to a new definition of planetary status, and there are two simple rules we can follow in order to determine this. First, anything that falls on the main sequence is a planet. Secondly, anything that falls off of the main sequence can still be a planet, so long as the area of the planetary H-R diagram it falls into has an analogous stellar H-R diagram location that is on a known main-sequence sub-branch. For instance, Mercury is in the transition from white dwarf to main sequence, and so would remain a planet. Jupiter, Saturn, Uranus, and Neptune all lie on the stellar H-R equivalent transitional phase between the main sequence and the red giant sequence, and thus would retain their planetary status also. However, asteroids fall in the bottom middle of the planetary H-R diagram, and ice dwarfs in the bottom right of the planetary H-R, and these two regions are *not* analogous to any subsequence on the stellar H-R, and thus these objects all fail as planets. Finally, we can actually synthesize these two rules down to one simple test for planetary status. That test is as follows: plot the candidate body on a planetary H-R diagram, and if it falls in the middle bottom or the middle right of the diagram without being on the main sequence, the candidate is not a planet. Otherwise, it is!

The Case For Life on Pluto

In retrospect, it seems ironic that a body that is cold and lifeless, devoid of air, and almost completely without water—one that is almost a textbook example of sterility—could, in fact, be the protector of, and one of the reasons for, all of the life currently known in our universe. And yet such is the fate of the Earth's moon.

The moon serves this function in many ways. The simplest of these is to function as Earth's deflector shield. Without the moon, the Earth would have been struck many more times by asteroids or cometary bodies. Thus, without the moon, there would have been many more impact-related mass extinctions in Earth's history, and what state life on our planet would currently be in could only be guessed at.

But the moon serves in other ways too. The moon also serves to stabilize the Earth's rotation, making sure that the variability in the Earth's axial tilt is never too extreme. This is fortuitous, because the sensitivity of Earth's seasons to the tilt of its axis is severe. Just a couple of degrees of tilt more or less can radically alter atmospheric heating, ocean currents, and it can eventually determine which areas of the Earth become hothouses or suffer ice ages. At first glance, this might not seem like such a big deal, as the Earth has been through ice ages before with life surviving intact, but what could not have been guessed at until recently is that just a five degree Celsius swing in temperature above or below the planetary norm of about twelve degrees Celsius (fifty-four degrees Fahrenheit) is enough to cause a mass extinction. Thus, life on

Earth might survive without the moderating influence of the moon, but intelligent life surely wouldn't.

Yet another effect of the moon's presence on life is its tidal pull, although we are not too sure of a direct linkage. Suffice it to say that many life forms display important cycles which are synchronized to the lunar month. We often ignore this aspect of the moon's presence because our current tides are perceived as rather benign. But what if they weren't? What would happen if the moon, instead of being 1/6 the size of the Earth, were half its size? What would the effects be if the moon had the same density as the Earth? And finally, what would happen if the moon, instead of orbiting at a comfortable 250,000 miles distance, were twenty times closer?

The calculations of the gravitational effects for these changes are relatively straightforward. The moon's radius would change from about 1,000 miles to about 2,000

miles (Earth's is 4,000), increasing the moon's volume by a factor of eight. The moon's density would also increase, going from somewhere around 3.3 gm/cm^3 to 5.5 gm/cm^3, increasing by a factor of about 1.7. Since mass equals density times volume, The moon's mass would increase somewhere around fourteenfold. But gravitation not only has to do with mass; it has to do with distance as well. Thus we must also take into account the fact that this more massive moon would be twenty times closer than before. This new distance factor alone would increase the gravitational pull of the moon by a staggering 400 times. When we put both the mass and distance factors together and multiply, we get a mind-boggling increase in the moon's gravitational effect of something on the order of 5,600 times.

One immediately wonders if this translates to tides going from two feet to two miles high in this hypothetical Earth/Moon system. Here we observe that since our oceans average about two miles deep, this would translate to most of our oceans being torn from their beds and thrown up on the land twice a day. There would be enough energy here to cause massive earthquakes. Quakes of ten to fourteen on the Richter scale would conceivably be a possibility. Any tectonic plate that did not rupture or rift due to such massive quakes on a regular cycle would certainly do so due to the enormous flexing of the Earth itself from such a massive moon being this close. The internal heat generated by the flexing would result in enormous upsurges in volcanism, and unheard-of amounts of greenhouse gases would be spewed out into the atmosphere as a result. The Earth would quickly become a hothouse. In fact, there is so much energy here that one wonders if the Earth wouldn't eventually be pulled apart, along with its massive companion, only to recoalesce into one massive body.

The first reaction to the scenario above would most likely be that it couldn't possibly happen, but what if we could give a scaled-down version of exactly that sort of system! What if a hypothetical system such as the one described above did, in fact, exist? As it turns out, a system like this does exist, and it is right in our own backyard. It is the Pluto/Charon system.

We don't propose that Charon and Pluto are tearing each other apart, but the facts remain: Charon is ½ the size of Pluto, it is of equal density, and it orbits as close as 12,000 miles to Pluto. [5] In other words, the relative effect which Charon has on Pluto is 5,600 times greater than the effect of our moon on the Earth. This is a tremendous source of energy.

One aspect of this gravitational tug that Pluto and Charon exert on each other is that it suggests a calculable mean density for Pluto of about 2 gm/cm^3. [6] Any density above 1.85 gm/cm^3 implies a makeup dominated by rocky materials. From this, it has been hypothesized that Pluto is made up of about 70% rock and 30% ices. This ratio is far above the fifty-fifty norm of bodies that form from the solar nebula. Pluto seems to consist of a rocky core (similar to those of Triton and Europa, among others) that is differentiated into either one or two layers of ice, and possibly volatiles.

In other words, significant amounts of water are thought to exist on Pluto, if only in a frozen state. Thus a second key ingredient for the formation of life is present on Pluto as well. The real question as far as Pluto's ice is concerned is if there is enough internal pressure at certain depths to heat the ices enough to convert them to water. Even if the critical pressure threshold is not reached, it still does not imply that liquid water does not exist on Pluto; however, the internal pressure gradients, when combined with the enormous gravitational flexing of the system, may well be enough to ensure the existence of liquid water under any set of circumstances. Thus, the gravitational field of the Pluto/Charon system provides not only an energy source, but possibly also hints at liquid water as well.

In addition to an energy source and water, Pluto may even have amino acids.

The evidence for this comes from the systematic occultations of Pluto by Charon between 1985 and 1990. For each occultation, astronomers carefully calculated the albedo of different portions of Pluto's surface during Charon's passage. The result of this was that they found there were significant coloration differences between regions on Pluto's surface. The inferences drawn from the coloration pattern was that some areas were light and some areas were dark or reddish. The light-colored areas seem to indicate the presence of nitrogen, methane, and carbon monoxide, along with traces of water. The dark or reddish areas are thought to be complex hydrocarbons formed by the ionization of gases by ultraviolet radiation, or the particles of the solar wind. [7] From the above evidence, it is an easy step to infer that some of these atoms are, in fact, amino acids (a prime candidate being adenine if the dark patches are more yellowish brown than dark or red). This is not so hard to accept when one remembers that amino acids are thought to have formed on other moons of the solar system, notably Europa and Titan, in a similar manner.

So, summing up, we have the distinct possibilities of viable energy sources, water, and amino acids on Pluto. These are the universally accepted three components necessary for life. Intriguing as it may seem, life may be as plausible on Pluto as it is on Europa, and for similar reasons. But before we declare this to be a real possibility, let's kick the tires a little. Just how much energy is present in the Pluto/Charon system, and just how much of an ocean could form?

Let us calculate the gravitational energy of the Pluto/Charon system as a percentage of that of the Earth/moon system. This percentage is a function of the four radii of the respective bodies and the density of each, as illustrated in the table below.

Body	Radius	Volume	Density	Mass
Earth	1.0000	1.0000	1.00	1.00000
Moon	.27	.0203	.61	.01230
Pluto	.1876	.0066	.36	.00240
Charon	.0919	.00077	.36	.00028

[8]

The relative mass product of the Earth/Moon system is .0123 = (1x.0123), while that of the Pluto/Charon system is .00000065 = (.0024 x .0002). Thus the combined mass contribution to total gravitational energy for the Earth/Moon system is 18,675 times as great *in absolute terms* as it is for the Pluto/Charon system. However, as previously noted, the relative distance squared factor for the Pluto/Charon system is a factor of 400 times that of the Earth/Moon system. Calculating this out in pure numbers, we get (.00000065 x 400)/.0123, or .00026/.0123, or .0214, for the ratio of the gravitational field of the Pluto/Charon system to the gravitational field of the Earth/Moon system That is to say, Pluto and Charon generate fully 2.1% of the gravitational energy of the Earth and Moon, despite being thousands of times less massive! This surprising result can be confirmed through the alternative method of calculating $F = Gm_1m_2/r^2$ for both the Earth/Moon, and Pluto/Charon systems and then dividing the results. In doing so, we see that that the G can be cancelled, as it occurs in both equations, leaving $F(pc) = M_pM_c/R(pc)^2 = $ (1.31 E22 x 1.5 E21) kg./19570^2 km. and $F(em) = M_eM_m/R(em)^2 = $ (5.97 E24 x 7.4 E22) kg./384400^2 km. = 5.11 E34/2.99 E36 = .0171 = 1.71%. This is on an absolute basis rather than on a relative one, so we must conclude that there is sufficient energy in the gravitational field of the Pluto/Charon system to support the possibility of life.

The amount of water that may be present is another matter. We have already seen how the relatively high density of Pluto (2 gm/cm^3) indicates a 70% rock composition, so a natural inference is that water ice can be approximately as thick as 30% of the planet's radius. Thus, water ice on Pluto may reach depths of about 210 miles (700x.3).

How much of this ocean may be in liquid form is a very interesting question. Part of the answer may depend on the pressure that the ice is under, under varying levels of the ice cap. On our own planet, a depth of thirty-two feet of water is equivalent to one atmosphere of pressure, thus a simple way to determine force in pounds per square inch is to take the ocean depth and multiply by .46875. A depth of one mile, therefore, represents a pressure of 2,475 pounds per square inch. On Pluto, neglecting the fact that the majority of any ice cap would be solid, the very bottom of a 210-mile-deep ice cap would face a pressure of 519,750 pounds per square inch! There must be enough pressure here to convert the ice cap to liquid water past a certain depth. But what is that depth? Since water is considered to be incompressible, one way to attack this problem is to assume linearity. That is, the temperature is added in a straight line as a function of pressure times a constant. Or using proportionality, $t2/t1$ must $= k \ x(1/(p1/p2.).$ So 273/40 Kelvin $= Kx$ $1/p1/p2 = .01 \ x \ (x \ / \ 15)$. Here we can see that even if the compressibility constant for ice were .01, ice would turn to water under about 10,237 lbs./sq. in. This is somewhere around 21,840 ft. in depth, or about 4.1 miles down. If the constant k were as low as .001, then ice would turn to water about forty-one miles down, and if K were as low as .0001, then ice would turn to water 410 miles down. This last figure, of course, is below our 210-mile limit, and thus it seems that unless the compressibility of the particular form of ice found on Pluto is greater than .0002, the presence of liquid water cannot be guaranteed from this form of analysis alone.

A better idea comes from the more constrained figures of an ice-field depth of 250 km with an implied pressure at that depth of .17 gigapascals. [9] The implication of these figures and of Pluto's internal configuration has led to recent estimates of internal heating to the levels of one hundred to two hundred Kelvin, or possibly a little more. [10] Currently, these temperatures are not tightly constrained. The real definitive study on this problem will probably have to wait until New Horizons arrives at Pluto. For now, however, the best we can say is that a little play in the upper value of the above figures would put some portion of Pluto's internal volume tantalizingly close to the 273 Kelvins necessary to support liquid water through internal pressurization.

Thus the real question becomes, is there enough gravitational energy in the tidal heating of the Pluto/Charon system to raise Pluto's internal temperature over some fraction of its volume by about seventy Kelvins? This question is notoriously difficult to answer at present. In fact, there seems to be no consensus on just how to approach this problem, as some researchers have variously modeled the problem as a gravitational tensor problem, a loaded spring body problem, and even a Navier-Stokes equation. The math in all of the above cases is horrendous. Yet, in at least one abstract, sub-ice oceans for all outer solar system bodies are estimated to form at depths of around 140 km for a system with a Cartesian geometry, and at depths of as little as 90 km for systems calculated under spherical geometry constraints. (Again, Einsteinian tensors are necessary here.) In addition, the surfaces of some moons are nonconductive, nonconvective, volatile ices (N_2 and CH_4), and thus act like insulators. This effect is extremely important, as it helps certain planets and moons to trap heat internally. Specifically mentioned in this regard are Triton and Pluto. Without the insulation effect of the volatile ices, liquid water seems to be reached on Triton at depths of 330 km. With the insulation effect, however, liquid water is reached at a depth of 200 km. [11] Given, the more exacting measurements of ice shelf depth above, the immediate implication of this finding is that between 120 and 150 miles below the surface, Pluto may well be harboring an ocean that is thirty miles deep!

Further evidence supporting ocean creation through tidal heating might be inferred from the atmospheric properties of the Pluto/Charon system. The salient point of this type of analysis is that Pluto's atmosphere periodically lies frozen on its surface. Yet Pluto is so small and has so little mass that any atmospheric gas that is energetic enough should eventually bleed off into space. This means that it would be difficult to account for the presence of any sort of an atmosphere unless Pluto generated it from within, or in other words, unless Pluto was volcanically active. Without a doubt, the presence of volcanism on Pluto would also mean liquid water. If this layer of liquid water was only two miles deep, life in the system would be theoretically possible. (Life on Earth developed in oceans averaging two miles in depth.)

One way to check on this issue is to calculate the minimum radius of a planet necessary to gravitationally bind a specific gas. If this radius is greater than Pluto's, then a gas atmosphere cannot exist on Pluto unless generated by an internal source. This calculation involves finding the equation for the escape velocity of the planet and setting it equal to the equation for the velocity of an ideal gas derived from the ideal gas law. The equation for the escape velocity of any planetary body is $V = (8/3xPixG*pxR^2)^{1/2}$

Here, pi is 3.14159, G is Newton's universal gravitational constant $(6.67 \times 10^{-11}$ nt-m^2/kg^2), p is the density of the planet (2.0 g/cm^3 in Pluto's case), and R is Pluto's radius (about 745 miles, or about 1137 km). We set this equal to the quantity $(3x Rx T/Mp)^{1/2}$, where R is the universal gas constant (8.315 joules/moles/Kelvins), T is the absolute temperature in Kelvins of the gas in the atmosphere, and Mp is the mass per mole of the gas, which if we assume is nitrogen since nitrogen is diatomic, is (2x14) gm/mole, or 28 x 10^{-3} kg/mole. Combining like terms, we get the equation Ve = $[(9xRxT)/(8xGxPixpxMp)]^{1/2}$. At this point, we need only two more minor adjustments and then we can simply plug in the terms and solve the equation. First, let's change density to the MKS system and define it as 2,000 kg/m^3 (since there are 1,000,000 cm/cubic meter, we are talking about 2,000,000 gm/cubic meter and dividing by 1,000, we get 2,000 kg/m^3), and second, let us define the atmospheric temperature (or surface temperature, since Pluto's atmosphere lies frozen on its surface) as 60 Kelvins. [12] The above equation now becomes Ve = $[(9x8.315x60)/(8x6.67x10^{-11}x3.14$ $159x2000x.028)]^{1/2}$. = $(47,830,210,563)^{1/2}$ = 218,701 meters. Thus, it would seem that since Pluto is so close to absolute zero, it could have a miniscule diameter and still gravitationally capture Nitrogen gas. (In this case, a radius of around 218.70 kilometers, or 137 miles, seems to do the trick.)

The real question then becomes, does this atmosphere necessarily have to freeze out? If the above calculations imply that any gas in and around Pluto should be lying frozen on its surface, then it becomes difficult to account for observations of Pluto that show a tenuous gas envelope encircling both Pluto and Charon at various times during the system's journey around the sun. At present, there is no consensus on this question.

However, even without an explicit answer, this form of inquiry begs a question that must be asked, and that is, given that the freezing point of Nitrogen is sixty Kelvins while that of methane is ninety Kelvins, how does Pluto ever get warm enough for methane to sublimate into its atmosphere in the first place?

We know from Pluto's orbit that there are times when it is not the most distant planet in the solar system. Periodically, Pluto moves inside the orbit of Neptune. Some exacting figures suggest Pluto, at its coldest, is around -235 degrees C, while at its warmest, Pluto appears to reach about -170 degrees C. [13] due to the above-mentioned high eccentricity of Pluto's orbit. Since K = C - 273, the implications of the above parameters are temperatures of 40 Kelvins (-419 F) at aphelion, and perhaps 103 Kelvins (-356 F) at perihelion. As an additional check on these figures, given Pluto's size, a very good approximation of what its surface temperature might be like during its closest orbit approach to the sun would be the temperature which exists on Triton, or perhaps a little bit warmer. Triton is the coldest moon in the solar system, experiencing temperatures of -400 degrees Fahrenheit. Both estimates of Pluto's temperature are fairly close at perihelion, and so we have some confidence that they are close to correct.

Even at closest approach, however, the activity of the sun is only enough to warm Pluto's atmosphere by forty to sixty K or so. The question remains, is the forty to sixty K temperature critical, and is it enough to unfreeze nitrogen, oxygen, or methane? The freezing point of Nitrogen is sixty-three Kelvins, that of methane is ninety Kelvins. Multiplying each by 1.8 and subtracting 459 from each should give the Fahrenheit equivalent. This gives Nitrogen's freezing point as -351 degrees (-210 C), and methane's as -297 degrees Fahrenheit (-182.5 C). [14] Thus, It seems from our cursory check that Nitrogen would be the only element with even a remote possibility of existing in a gaseous form on Pluto, and just barely, in this case making the threshold by as little as five degrees Fahrenheit. However, the methane is clearly a problem. Even under the most generous of temperature estimates, the implications are that Pluto is fifty-nine degrees Fahrenheit too cold for methane to exist in a gaseous form. It is entirely unclear if Pluto ever gets warm enough to sublimate methane, and indeed, the most recent measurements of Pluto's atmospheric temperature have shown an upper limit of only one hundred Kelvins or so, which is consistent with the highest temperature estimate at perihelion, given above. Granted, methane has a tremendous greenhouse effect, and an atmosphere that reaches something like 3% methane content may maintain enough internal heat to sublimate methane, but this still begs the question of why Pluto would not have a much thicker atmosphere than we see now if methane were continually sublimated, and it also begs the question of just what event in Pluto's past kick-started the sixty-degree F increase in temperature necessary to put that methane in the atmosphere in the first place.

It would seem that a much better theory is that perihelion approach sublimates enough Nitrogen to mitigate the anti-greenhouse effect known to exist on Pluto and to mitigate the blanket effect of the nonconvective volatile ices to a certain extent, thus allowing heat trapped in the interior to percolate up to the surface. If this were the case, we would expect Pluto to actually get warmer, not colder, as it moved away from perihelion. This is exactly what we are seeing. Recent calculations show that Pluto's atmospheric pressure has tripled, and its surface temperature has gone up three degrees Fahrenheit in the fourteen years since perihelion. [15]

Given the above arguments, it would seem that one of the few, if not the only, other possibilities left open to us if Pluto never gets over -356 Fahrenheit is that it has to somehow be volcanically active. The new Pluto/Kuiper Belt mission may get a surprise during flyby of active Nitrogen geysers on Pluto that are similar to what we see on Triton, or perhaps there will even be an active volcano like the ones on Io.

With volcanic activity of some sort becoming increasingly possible, the presence of liquid water is becoming more probable. So we conclude that life around underwater vents is at least a possibility on Pluto. This is an important conclusion in and of itself. If life becomes theoretically possible, even on the most extreme and most distant example possible, then the possibility of life in the universe must go up dramatically. The search for extraterrestrial life would then transform to not just a simple exercise of counting planets and moons and applying a percentage of probability for life to exist to these numbers, but to a search where every system would have to stand and be analyzed on its own merits no matter how extreme the system.

This is not the only lesson that the Pluto/Charon system has to offer.

The relatively large size of Charon compared to Pluto, and its extreme closeness, is a feature that makes this a microcosm by which we can judge our own system. Now we have come full circle, because while the example with which we opened our examination seemed impossible for the Earth and Moon today, it is, in fact, very similar to the situation that existed 3.8 billion years ago, right after the moon first formed. Back then the moon orbited only 16,000 miles away from the Earth. While the moon was as close as we originally hypothesized, its gravitational effects were not increased by 5,600 times, which was used only as a relative example of what would happen if the moon was the same size and density compared to Earth as Charon is compared to Pluto, but by a factor of 256! This would still translate to tides 500+ feet high on the early Earth. The amount of gravitational energy here is still enormous, and it may have been a key ingredient for the development of life on the early Earth. It may be no coincidence that the earliest known evidence of life is also 3.8 billion years old.

This may also answer an interesting question. From 4.5 to about 4 billion years ago, no life could have existed on Earth. This is because asteroids and protoplanets perpetually bombarded Earth. After the original bombardment ceased, however, the planet became remarkably cold. This may seem a strange conclusion until we remember that the sun burned with 30% less output 4 billion years ago than it does today. This would be a sufficient decrease to leave water frozen on Earth's surface. So where did the liquid water which life first formed in come from? The obvious answer is that the proximity of the moon pumped enough gravitational energy into the early Earth to exactly compensate for the lack of solar energy. It would be a fascinating theory indeed if we could prove that as the solar energy input increased through time, the gravitational energy of the moon decreased, by the conservation of angular momentum swinging the Moon's orbit outwards, in exact proportion to the solar energy increase. In other words, the total energy input of the Sun and moon remained constant over Earth's history. One other fascinating idea springs from this: no moon . . . no liquid water . . . no life on Earth. From this argument, we can see the importance of the New Horizons mission. In exploring the Pluto/Charon system, we may be seeing a small-scale replay of the first formation of life on Earth. This alone makes the price of the trip worth it.

What Can Pluto Tell Us About Our Own Solar System?

Of all the facts and figures that we have presented so far, perhaps the most anomalous is the ice and rock content of Pluto. The high rock content of Pluto marks the composition of the entire system as different from the primordial solar nebula from which it presumably sprang. Thus we now ask the question, where did Pluto originally come from if not the solar system?

Perhaps the easy answer to this question is to simply state that Pluto coalesced as a major body either in the Kuiper Belt or in the Oort Cloud. This would mark Pluto as a capture. The other member of Pluto's system is probably also a capture. The extreme tilt of Pluto's spin axis (122 degrees) makes this likely; as such an extreme tilt is a known signature of moon formation through collision. In this case, Charon would have formed out of debris captured by Pluto during a collision with some unknown body. This is not unknown in the solar system. Uranus also has an extreme tilt (ninety-seven degrees), and at least one of its moons, Miranda, is thought to have formed from a collision of a large body with Uranus. Even our own Moon is thought to have formed when a Mars-size object hit the proto-Earth. We tend to analyze these events in and of themselves, but when put together like this, the implications are staggering, as fully three of the original nine planets of our solar system suffered some sort of titanic collision with planet-sized bodies in the past. Even more remarkable is the fact that studies of impact rates on our own moon and other moons of the solar system lead to the conclusion that all of these events date back to the same point in time—around 3.8 billion years ago. (This is also the time when life may have started on Earth). Something very strange and unbelievably cataclysmic happened to the inner solar system during this time period. So, if Pluto is a Kuiper Belt capture, the question now becomes, what unknown catastrophe sparked this bombardment and Pluto's eventual capture 3.8 billion years ago? Whatever this event was, its occurrence must have been extremely rare, as this type of event has not repeated itself after nearly 4 billion years.

The above arguments reveal that in some respects the capture hypothesis for Pluto is unsatisfying, because it not only leads to a catastrophic event for which we currently have no good model, but also because it simply pushes the question back a level without really answering it. This should certainly trigger a follow-up; if Pluto's composition is like that of the Oort Cloud or the Kuiper Belt, then why is the Oort Cloud or Kuiper Belt composition so different from that of the primordial solar nebula?

Considering the fact that the trillion or so cometary bodies and ice dwarfs roaming the outer solar system should have condensed from the same nebula as did the sun and the planets, any claim of differing compositions between the two groups leads to a logical contradiction. In fact, if held on to as a hypothesis, the direct implication of differing compositions is that the inner and outer portions of the solar system, in fact, evolved in different locations.

An examination of some of the rarer attributes of our inner solar system may provide clues that will either verify or falsify the above hypothesis. As a whole, conservatism in science has led to the conclusion that physical processes are the same in all corners of the universe (Poincare invariance), and that a natural consequence of this viewpoint is that our solar system is really no different than any other. Sometimes, this orthodoxy does not serve science well, however, as the most important advances in science often derive from seeing through to the exceptions to the rule and by not letting the gradualism of the rule suppress the laws of the extreme that can be garnered by analyzing the stranger cases. Such is the case with our solar system, and specifically our sun. The plain truth about our sun is that it is different—very different! Specifically, our sun appears to be the most metal-enriched star that we know of, possibly in the whole universe. Our Sun is also the only star that we can prove supports life at the current moment. These two facts juxtaposed cannot be a coincidence.

The extreme metal enrichment of our Sun is extremely unusual for stars located in spiral galaxies. Almost all of the heavily metal-enriched stars that we know of seem to form either in Dwarf Irregular Galaxies (DIGS), or Dwarf Elliptical Galaxies (DEGS). This fact, combined with the implications of differing compositions between the inner and outer solar system, leads us now to propose our hypothetical bombshell: The Sun is not a member of the Milky Way Galaxy! The Sun, in fact, comes from the outside, and most probably formed in one of the DIGS or DEGS in our local group.

The next question obviously is, if the Sun is not a member of the Milky Way, then how did it get here? The only reasonable answer to this seems to be that the Milky Way collided with this DIG or DEG around 3.8 billion years ago. This collision has to have been the seminal event in the formation of our current solar system.

We say current because our hypothesis may now shed some light on some of the more curious facts gathered in relation to the composition of our solar system, specifically the chemical compositions of Uranus and Neptune. It has long been known that the composition of these two bodies differs from that of the solar nebula. As an example, the two gas giants, Jupiter and Saturn, do show compositions similar to our solar nebula; they are both about 90–93% hydrogen, and 7–10% helium. [16] Uranus and Neptune, on the other hand, do not display this signature. The composition of Uranus and Neptune tends to run something on the order of 80% hydrogen, 18% helium, and 1–2% methane. [17] Astronomers have long been hard pressed to explain a mechanism for this difference, but given the above hypothesis as to the origin of our Sun, we can now drop a second hypothetical bombshell: Just as the Sun is not a member of the Milky Way Galaxy, Uranus, Neptune, and Pluto are not members of our solar system! They, too, come from the outside.

Here, we ask a question similar to that of the one we asked about our Sun, and that is, if Uranus, Neptune, and Pluto are not members of our solar system, then how did they get here? Just as the most logical answer to the question of how the Sun got here seems to be that it formed in a collision with another galaxy, the answer to how Uranus, Neptune, and

Pluto got here seems to be that they formed because of a collision with another solar system! When the DIG containing our Sun's original solar system collided with the Milky Way 3.8 billion years ago, it must have passed through, or very close to, another solar system already resident in the Milky Way itself.

This conclusion leads one to speculate as to exactly what our Sun's original solar system looked like. Specifically, if Uranus, Neptune, and Pluto are here, where are the original outer planets of our solar system? The logical answer seems to be that if Uranus, Neptune, and Pluto are here, then the original planets of our solar system must be out there! In other words, a crossover occurred!

This may not be as strange as it sounds, as it tends to answer a question about the multiple planetary collisions outlined above. That question, roughly put for the case of the Earth, might run something like, if the Moon was formed out of the collision debris of the Earth with a Mars-like object, then where is that object? Here we can only think of three possibilities: ejection, destruction, and collection. The total destruction of planet-size bodies seems improbable. In the case of the Earth/Moon system, there is insufficient mass to account for the total mass of an Earth/Mars-like system. No matter how violent an impact, one would expect that something would be left over, either a whole body or sizable fragments with a relative mass on the order of the difference between that of Mars and the Moon. It is equally clear that this missing mass is nowhere to be seen in the area. Adding to this conundrum is the fact that the estimated impact was known not to be violent; it was basically a gravitational grazing blow, and therefore, a Mars-sized body should be present in the solar system. This just makes sense. If the Sun's gravitational field is strong enough to capture Mars in an orbit 140,000,000 miles further out than Earth, then it certainly must be able to capture a slowly moving Mars-sized body much closer in. Of course, one solution might be that the Mars-sized body that hit the ancient Earth is, in fact, the body we now call Mars, but at present there seems to be no viable test of this hypothesis. This leaves ejection as the only viable alternative. A pure change in trajectory could lead to the ejection of this body, but even here we suspect that the Mars-sized body could not have been ejected any further out than the Oort cloud, and even though the existence of dozens, if not hundreds, of Mars-sized or larger bodies in the Oort cloud has been theoretically conceded, one would suspect the presence of such a body would cause periodic cometary bombardments—something we do not have evidence for at the present. Perhaps a more elegant and simpler solution is that the hypothesized Mars-sized object that is missing was ejected, but ejected as a capture by the solar system that was originally resident in the Milky Way when it collided with the Sun's DIG.

We might even be able to estimate the attributes of the original resident solar system if Uranus and Neptune were the gas giants of that system and fulfilled the same role as Jupiter and Saturn do in our solar system. This would imply a solar system with constituent planets possessing radii with a constant signature of 2.7 to 4 times less than, and relative volumes averaging eight times less than, the analogous planets in our solar system. This leads us to suspect that the gravitational energy needed to bind such planets is much less than that in our own system. The natural implications of these considerations are that the solar system that was originally resident had a red dwarf star for its sun.

This is almost certainly not the first time a hypothesis along these lines has been proposed. What is different about our hypothesis is that the violence of the collision and the reshuffling of the components of the constituent solar systems is taken to an extreme. The major test of this hypothesis, of course, is whether or not a successful computer model can be developed to show the likelihood of the above scenario without any major drawbacks. Critical in this regard will be the closest distance of approach between the two stars, the velocity of both solar systems through space, and the angle of attack, for lack of a better phrase.

It has been shown through other models that a red dwarf could have passed by our solar system at as close as 14 billion miles (a little less than 160 AU, or four times farther out than Pluto) without being captured, and at the same time those models explain some of the more mysterious features of our solar system, such as why the Kuiper Belt shows such an abrupt cutoff, and how a planet like Sedna might have been captured during the collision. [18] The above hypothesis, however, almost certainly calls for this approach to be a lot closer than 14 billion miles. If the red dwarf's solar system extended out into space to the same degree as the Sun's original system, then we could expect there to be about 7 billion miles in separation between the two stars while the outermost members of both solar systems would just begin to come into contact. Since the crossover hypothesis calls for the exchange of Uranus and Neptune-like bodies, this distance may even be more severely restricted to about 5 billion miles (1.7 billion to Uranus, 1.7 billion in overlapping orbits between the two systems, and 1.7 billion more from the red dwarf star to its analogous Uranus-like body).

The overriding question for such a short semimajor axis is, would the red dwarf be gravitationally captured by our Sun in the event of such a close approach? This is where the mass of our Sun and the velocity of the red dwarf become critical. The mass of the Sun controls the escape velocity function relative to the Sun's gravity at varying distances throughout our solar system. For our solar system, a good approximation of this function seems to be $41.8/(AU)^{1/2}$. [19] This function gives escape velocities in the vicinity of Uranus of somewhere around 9.6 km/sec. At a distance of 5 billion, miles the escape velocity with respect to the Sun's gravity seems to fall to around 5 km/sec. Thus, the red dwarf star would have to be moving at a velocity between 5 and 10 km/sec. with respect to background stars in order not to become gravitationally bound to our solar system.

An interesting comparison with these figures comes about when the orbital velocities of any captured objects starting at 5 billion miles are calculated using Kepler's third law divided into 2pi(AU). Evidently, the orbital velocity of a captured object at about 54 AU would be something on the order of 4 km/sec. This figure is extremely close to the escape velocity from the Sun at that same distance, and serves to underscore that any nudge, jolt, or small outside force applied to any object in such an orbit would likely eject that object from the solar system. This makes any capture past certain distances extremely unlikely, and even if a capture did by chance occur, such objects are likely to be ejected after only a few million years at best.

Thus, it appears possible for a red dwarf star to pass within 5 billion miles of our solar system and not get captured in the process. This, in turn, makes it seem theoretically possible at least for the trailing edges of two different solar systems to interact over small time scales. Like two hurricanes rotating in opposite directions, the sheer forces generated in the zone of intersection would be enormous. The immediate consequences would be felt in the Oort clouds of both solar systems. Extending out perhaps a light year from both systems, system velocities on the order of 20 km/sec would imply bombardments of both systems by trillions of comets lasting something on the order of 15,000 years.

Critical in this regard is what can be called, for lack of a better term, the angle of attack for the two solar systems. Neither system necessarily has to lose half of its spherical shells of Oort cloud comets if one or the other solar systems "obliques" the other in the plane of the ecliptic. Along similar lines, the Z coordinates of both solar systems are also critical. If the ecliptic planes of both systems are misaligned along the z-axis, a second oblique angle of attack is created. Where several obliques exist at once, a small fraction of both Oort cloud shells would be lost, and that is all. Such a strike might also explain why Pluto is so highly inclined to our ecliptic plane. This might also explain the presence of so many Eris- and Pluto-size bodies in our own Oort cloud, especially if the Red Dwarf system was stripped of its outer shell due to the mass dominance of our own Sun.

The dynamics of such a collision also leads one to suspect whether or not retrograde planetary rotation is a signature of original residence in the red dwarf star system. If so, then the origination of the planet Venus might become an open question. This may not be as out of the question as it sounds, as it has always been a mystery as to why two almost identical Earth-size planets would form so close to each other in the original solar nebula. With this consideration, it could well be that Venus, Mars, Uranus, Neptune, and Pluto are all crossover objects, and the only original members of our own solar system that are still left are Mercury, Earth, Jupiter, and Saturn.

An immediate test of such an extreme hypothesis would be to identify the DIG that may have collided with the Milky Way and see if the ephemeris calculations for both galaxies indicate that that collision occurred 3.8 billion years ago. One possibility is that two suspects could be the Large Magellanic and Small Magellanic Clouds, as they are thought to be currently in collision with the Milky Way. However, another even more promising suspect is far closer to home, and that is the Canis Major Dwarf Galaxy. Some components of the Canis Major Dwarf Galaxy are as close as 5,000 light years to our own, and if other components are smeared out along the edges of our Galaxy in an arc, then it would be fascinating indeed if these attributes pointed to the Canis Major Dwarf Galaxy as our Sun's original home. If the ephemeris is wrong for the Canis Major Dwarf Galaxy possibility, however, then the most promising candidate for the original DIG is a yet-to-be-discovered dwarf galaxy, possibly one that is obscured by the center of our own Galaxy, or one that has been almost completely devoured and has left behind no sizable fraction of itself.

Perhaps even more interesting than the physical parameters of such a model are the philosophical implications behind the model. If the model is correct, and Pluto is a capture from another solar system, then what does this do to the current debate over planetary status for Pluto? For example, do we now have to decide whether or not what is now called Pluto was truly a planet in its own solar system? The natural follow up of this question is even more interesting, what happens, for instance, if we decide Pluto was a planet in its own red dwarf system, but rule any object that is a capture can't be a planet, essentially stating that a body has to originate in its own system to be called a planet, even though it was once a planet elsewhere? All of this would fly in the face of the fact that Pluto fits so 'nicely' into our current solar system that it 'looks' resident. The facts surrounding Pluto, its nice fit and external origin, then speak to the methodology of defining planet through a planetary H-R approach. A real interesting question here is if the external origin invalidates that approach for a collision generated system, or if it in fact strengthens it. This is implicit when we consider that the 'fit ' speaks to an underlying process of planet generation in both systems that produce systems along a tightly constrained set of variables, so that the fit is not noticeable. In any event the author would await the construction of a mathematical model of just such a collision with great interest, because the author is more curious about what can be learned from the radical nature of the approach outlined above than he is in starting a firestorm. Whatever is gleaned from such a model, it is hoped that New Horizons will turn out to be as stunningly successful a mission as the MER, or Cassini, allowing us to test theory with observation.

A mega – successful New Horizons mission would be as good as it gets, for if Pluto does turn out to be an outsider, then we begin to develop some insight into the Drake Equation. There is an inkling here, that every critical step in the formation of intelligent life on our own planet is impact related, and thus, life and especially intelligence are the end products of the extreme stresses and selection pressures generated by multiple catastrophes. Consider the following, the universe started in the greatest explosion of all time, the big bang. The first term in the Drake equation, the number of stars in our galaxy is now highly influenced by the number of catastrophic collisions with dwarf galaxies it has had in the past. Bacterial life on our planet seems to start right at the point in time where the Earth collides with a Mars sized object, forming our moon. Multicellular life appears at the end of a Snowball Earth like event, just before the Cambrian explosion. The snowball can be explained by the cosmic winter of a colossal impact. Phylum diversification can be explained from an apparent quickening of tectonic drift by a factor of twenty during the Cambrian, where the accelerated drift is best explained as a change in viscosity of the Earth's Mantle, caused by an unknown upsurge in heating in the Earth's core. In turn the heating of the Earths core implies the burning of a different fuel, perhaps the remains of the colossal impactor that caused the snowball. Warm bloodedness first appears right after the greatest extinction of them all, the Permian-Triassic extinction. We know for certain there was a colossal impact during this time period, leaving a recently discovered 300-mile wide crater under the 2-mile thick Antarctic ice sheet. Another colossal impact wipes out the Dinosaurs, insuring that intelligence evolves from a Mammalian lineage rather than a Troodon like one. And, finally, right at the critical Pliocene – Pleistocene juncture where proto hominids change from a small ape like creature to a six foot tall highly intelligent predator, at the very least we know from Ngorogorno in Africa, and Yellowstone in America that there were several super volcanoes during this time frame. It seems as if virtually every term in the Drake equation depends upon some sort of impact probability, or that perhaps a new form of the equation can be written based upon impacts, or other disasters. In any event, the above considerations would earmark multiple impacts as critical to life's development, and so tend to modify the Rare Earth hypothesis. Even here however, the existence of a large Jupiter – like world, screening the inner solar system from comet impacts should not be construed as likely to reduce the probability of life in a given solar system, because the highest possible impact velocities associated with comets seems to be right around 144,000 mph, a figure that seems to be fairly close to the escape velocity of Jupiter. Thus, far fewer bullets would get through, it is true, but the ones that did get through would be super-bullets. This would limit the number of impact related extinctions on Earth, but tend to intensify impact related extinctions into great extinctions. It is this intensification process that is the crucible in which intelligent life is synthesized.

In looking back over the preceding arguments we have come a long way, from considering planetary composition, all the way to the origins of life itself, and the planetary science associated with just how common life is thought to be in the Universe. While some of the suggestions above are necessarily radical, there also seems to be a logical flow to the arguments presented. Thus, our suggestions on the surface of it are at least rational, even if they have never been put forward before, something of which we aren't even sure. If some of the above hypotheses turn out to be confirmed through observation it would be remarkable, but even if they didn't we would still gain immediate insight into the formation of our solar system and its constituents by falsifying the catastrophist approach, leading to a more gradualist model of solar system formation. In all of this, perhaps the most amazing thing of all is that the entire trip was generated by thinking through the implications of the ice – to – rock ratio of a tiny oddball ***planet*** known as Pluto!

Endnotes

[1]   A newly published, and magnificent source for all physical data on the Pluto/Charon system is *The Encyclopedia of The Solar System*, McFadden, Weisman, and Johnson, Second Edition, Academic Press, Elsevier Inc., San Diego, California, 2007, pp. 541- 557. Pluto's surface properties and description appear in Section IV of Chapter 29.

[2]   Many sites on the web list the physical characteristics of Trans-Neptunian Objects, just one of these is http://www.johnstonsarchive.net/astro/tnos. html

[3]   http://www.sciencedaily.com/releases/2006/02/060206233911.htm

[4]   *The Encyclopedia of The Solar System*, McFadden, Weisman, and Johnson, Second Edition, Academic Press, Elsevier Inc., San Diego, California, 2007, pp. 541- 557. Pluto's orbit and spin characteristics are in Section II of Chapter 29.

[5]   http://www.solarviews.com/eng/pluto.htm

[6]   Ibid, or *The Encyclopedia of The Solar System*, McFadden, Weisman, and Johnson, Second Edition, Academic Press, Elsevier Inc., San Diego, California, 2007, pp. 541- 557. Pluto's interior and bulk composition are mentioned prominently in Section V of Chapeter 29, although Pluto's density is mentioned throughout the chapter.

[7]   Web site http://www.space.com \New Pluto Map Colors a Dark World.htm

[8]   These characteristics can be found on any web encyclopedia, or even inferred from a simple source such as *The World Almanac and Book of Facts 2007*, World Almanac Education Group, Inc., A WRC Media Company, New York, New York. Pp. 333- 337

[9]   *The Encyclopedia of The Solar System*, McFadden, Weisman, and Johnson, Second Edition, Academic Press, Elsevier Inc., San Diego, California, 2007, pp. 541- 557. Pluto's interior and bulk composition are mentioned in Section V of Chapeter 29,

[10]   Ibid

[11]   A. G. Fairen and J. Ruiz. Seas Under Ice. *Lunar and Planetary Science* XXXIV 2003.For example Http://www.lpi.usra.edu/meetings/lpsc2003.pdf

[12]   http://www.solarviews.com/eng/pluto.htm

[13]   Ibid

[14]   http://curious.astro.cornell.edu/, among hundreds of other sites

[15]   Http://www.space.com/scienceastronomy/pluto_warming_021009.html

[16]   Again, even as simple a source as *The World Almanac and Book of Facts 2007,* World Almanac Education Group, Inc., A WRC Media Company, New York, New York. Pp. 335-337 will suffice for this general info, as well as thousands of web sites.

[17]   Ibid

[18]   Did our Sun capture alien worlds? Dec 04 ,Space & Earth science http://pda.physorg.com/lofi-news-solar-system bromley_2193.html

[19]   This information can be inferred from a simple table, a list of escape velocities with respect to the Sun's gravity in Earth Velocities at even a simple site such as http://www.answers.com/topic/escape-velocity? Cat=technology, the mathematical equation inferred from putting the table contents into something as simple as an Excel spreadsheet is probably close enough for our purposes.

www.ingramcontent.com/pod-product-compliance
Lightning Source LLC
Chambersburg PA
CBHW021939170526
45157CB00005B/2347